Hardscaping

with Decorative Concrete

Tina Skinner

Schiffer Publishing Ltd
4880 Lower Valley Road Atglen, Pennsylvania 19310

Other Schiffer Books by Tina Skinner

Building Outdoor Environments with Retaining Walls.
Built to Last: A Showcase of Concrete Homes.
Creative Patios.
Hot Tubs & Spas.
Master Built Pools & Patios: An Inspiring Portfolio of Design Ideas.
Outdoor Rooms for Fresh-Air Kitchens and Living Areas.
Walkways & Drives: Design Ideas for Making Grand Entrances.

Other Schiffer Books on Related Subjects

1000 Fences and Gates. Jo Cryder
Arts & Crafts Designs for the Home. Douglas Congdon-Martin
Backyards and Boulevards. David R. Smith & the Interlocking Concrete Pavement Institute
Decorating with Concreted Ourdoors: Driveways, Paths & Patios, Pool Decks, & More. Tina Skinner & ConcreteNetwork.com
Decorating with Concrete Indoors: Fireplaces, Floors, Countertops, & More. Tina Skinner & the ConcreteNetwork.com
Designs for Garden Paths. Heidi Howcroft
Outdoor Rooms: Ideas for Fresh-Air Kitchens and Living Areas. Tina Skinner & Melissa Cardona
Patios, Driveways, and Plazas: The Pattern Language of Concrete Pavers. David R. Smith & the Interlocking Concrete Pavement Institue
Paver Projects: Designs for Amazing Outdoor Environments. Melissa Cardona, PaverModule
Retaining Walls. National Concrete Masonry Association, Tina Skinner.

Library of Congress Control Number: 2006937184

Designed by John P. Cheek
Cover design by Bruce Waters
Type set in Futura Hv BT/Zurich BT

ISBN: 0-7643-2598-1
Printed in China

Published by Schiffer Publishing Ltd.
4880 Lower Valley Road
Atglen, PA 19310
Phone: (610) 593-1777; Fax: (610) 593-2002
E-mail: Info@schifferbooks.com

For the largest selection of fine reference books on this and related subjects, please visit our web site at **www.schifferbooks.com**
We are always looking for people to write books on new and related subjects. If you have an idea for a book please contact us at the above address.

This book may be purchased from the publisher.
Include $3.95 for shipping.
Please try your bookstore first.
You may write for a free catalog.

In Europe, Schiffer books are distributed by
Bushwood Books
6 Marksbury Ave.
Kew Gardens
Surrey TW9 4JF England
Phone: 44 (0) 20 8392-8585; Fax: 44 (0) 20 8392-9876
E-mail: info@bushwoodbooks.co.uk
Website: www.bushwoodbooks.co.uk
Free postage in the U.K., Europe; air mail at cost.

Contents

Introduction

Today's consumers have a choice of almost any look they can imagine or find around the world, created in the easily accessible medium of concrete. The appearance of exotic stone, marble, brick, and worn cobble paving can all be recreated in decorative concrete. Prices are very competitive, and maintenance, safety, and durability are often vastly improved when traditional materials are duplicated in concrete. Beyond the traditional looks, today's technology is enabling craftsmen and artists to experiment and explore the ever-expanding uses for concrete. The variety of new permanent colors for concrete is infinite and the textures limited only by imagination.

Images in this book were drawn from the archives of the Bomanite Corporation. The leader in the architectural concrete paving industry for over fifty years, Bomanite has a network of over 250 specially trained and licensed franchise partners in more than seventy countries. This highly skilled and experienced network of concrete craftsmen has contributed to the hundreds of creative ideas represented in this book.

The Bomanite process was the first cast-in-place, colored, textured, and imprinted architectural concrete paving, developed in the middle 1950s in the Monterey, California, area. The originator and inventor, Brad Bowman, refined and expanded the process until 1970 when Bomanite Corporation was established in its present form. From its Palo Alto, California, offices, Bomanite Corporation began expanding its reach and experience by establishing franchises around the United States and other countries around the world. The company moved its corporate offices to Madera, California, in 1990.

In addition to its network of franchise partners, Bomanite Corporation also manufactures many of the components used in the Bomanite process, including dry shake color hardener, integral color, release agents, imprinting tools, fibers, and various other products for the installation of architectural concrete paving and flooring.

This book takes you on a visual journey through a great variety of hardscaping projects, from patios and porches, to elaborate pool environments. There are hundreds of ideas for adding curb appeal to a home via a spruced up entry and drive.

An unlimited palette of color and texture possibilities allow free rein to today's landscaper.

Driveways

A faux brick border around the driveway coordinates with the brick walls of the house.

A sandstone-colored expanse of flagstone-stamped driveway is a fitting complement to this handsome home.

A grand entry beckons with a nautilus border and European fan pattern.

Fresh from a recent rain, an entranceway is distinguished by color variations.

In the past, stamped concrete was extremely popular in commercial projects. Today the technology is increasingly tapped for high-end residential projects.

Adding an organic touch, as well as design, texture, and drainage, grass has been allowed to grow in the spaces between the concrete tiles of this driveway.

A grove of palm trees forms an oasis in the middle of the driveway created in warm tones and a random-stone pattern.

A carefully crafted patio incorporates a variety of stains and textures sure to fool amateurs into thinking this surface was created from carved marble, brick, and stone. The compass rose makes an excellent centerpiece, framed by a sculpted flowerbed ready for planting.

Stamped concrete offers a smooth surface, though variations in pattern and the visual effect of a brick or cobblestone surface can help to slow drivers down.

The sandstone color of this driveway works well with the red ceramic tiles and white stucco walls of a California complex.

Like a good carpet, a concrete pad can be etched and colored to decorate a driveway.

A random stone stamp pattern is often used by those who want their concrete surface to have a more natural-looking appearance.

A concrete pattern and palette can be chosen to match any home.

Carefully colored "flagstones" create the illusion of a real slate surface.

Because it is so versatile, concrete can be stained and textured to achieve almost any effect.

The smooth surface of a scored concrete walkway adds a level of safety for pedestrians, and makes snow and leaf removal easier.

Stamped concrete perfectly mimics brick.

A European fan pattern works well within a handsome circular driveway.

The gray of an ashlar-slate pattern offers cool contrast to the red-tone bricks found on the house and garage.

One common trick in concrete design is to leave the concrete between the stamp patterns its original color, creating the illusion of mortared stone.

A European fan pattern is an excellent complement to this Spanish-style building.

A parking area in front of this house has been carefully divided into warm-colored blocks, minimizing the effect of the concrete expanse.

In this driveway, the concrete was stamped to resemble glazed tiles like the kind used in Spanish style architecture.

Dark and vivid, a driveway stamped in a running bond pattern invites a closer look.

Another example of a random stone pattern.

Three angles emphasize the extensive entryway landscaping around a log home.

The border stamp and tone was slightly varied from that of the walkway.

Using a pattern of smaller squares in border stripes that zone larger sections of this entryway area, the builder created subtle visual variety.

Entryways

A European fan pattern complements the grand style of this entryway.

The concrete pattern has been continued right up to the edge of the steps in order to further simulate the look of pavers.

When concrete is stamped in an alternating pattern like this, it is called a basket weave, dating back to when such patterns were laboriously laid out in brick.

Colored tiles form a simple mosaic pattern reminiscent of patterns seen on indoor floor tiles.

A concrete entry stair and patio was created with various textures, and a built-in trench ready to fill with soil and foliage.

A white border defines a patio extension to this attractive entryway, the new "stone" color blending nicely with the terracotta tiles of the porch.

To decorate this walkway, builders chose colors that complemented the building.

Smoothed and polished, this lightly stained concrete underlines the rough textures and colors of the stone columns.

Using stenciling and staining it is possible to add decoration to even preexisting concrete structures.

Sandstone texture adds definition and color to an entrance, matching the pattern found on the driveway.

The borders and "grouting" were left the natural concrete color, heightening the beauty of the stained "tiles."

In the past, ashlar stone was measured, cut, and dressed by hand. Today, stamped concrete can achieve the dignified look in a fraction of the time with a fraction of the effort.

Not just for getting from the street to the house, concrete walkways are fast becoming more and more elaborate and decorative as important elements in a home's curb appeal.

Lights set underneath the lip of each step add an inviting decorative element as well as making them easier to navigate in the dark.

Instead of concrete or stone, a more organic edging was used in this walkway.

Faux tile insets create an added layer of luxury in a Romanesque entry court.

An obelisk and white concrete spheres act as automobile barriers, but also serve as lovely focal points on a beautiful starburst pattern in three tones of stained concrete.

Walkways

A random slate pattern implies relaxation and calm and encourages travelers to take their time along the path.

The color and pattern of the concrete for this walkway was chosen by the homeowner to complement the architectural style of his house.

Stamps, trowels, and skilled labor turned this staircase into a "stone"-capped pathway.

Waves of blue on dual ramps emphasize the gentle slope of pedestrian and handicap-accessible walkways.

Protruding edges imitate stacked stone in a staircase carefully handcrafted to mimic an aged walkway.

Bull-nosed steps complement the style of the nearby wall caps and the random slate pattern of the walk, adding detail to a very smooth and natural look.

Slate texture combines with ivy and a wrought iron gate and fence to create a classic looking pathway.

Rough-edged steps complement the random pattern stamped on the walkway.

A narrow border marks the circumference of a flowerbed.

A garden path passes small patios and focal areas on the journey to the pool.

A random slate pattern adds drama to the entryway of this tropical home.

In choosing colors, homeowners tend to be drawn to those supplied by nature in local stone. Concrete is easily adaptable.

A cobblestone-patterned walkway complements the fieldstone walls and slate roof of this storybook style house.

Ashlar slate meets herringbone brick as the front porch merges with the front walkway.

A small flowerbed divides a walkway and adds visual interest to the passageway.

A series of rounded steps spill down an incline, stamped with a random stone pattern. When the landscaping catches up, this walkway will appear to be a natural feature of the forest setting.

Stained and stamped concrete is durable, even in areas that require snow removal.

Another example of a basket weave pattern, this one featuring four rectangular blocks placed around a smaller square one.

If not for the risers between the steps, it would be very difficult to tell that this walkway is made out of concrete.

The tumbling block pattern heightens the sense of three dimensions in a walkway and patio.

A natural edge was created for this walkway by removing the partial wet concrete "stones" after imprinting.

A random effect was created by using the imprint mold to determine the edges.

A medley of colors and textures gives this patio an unique appearance.

Porches and Patios

Creating different levels and borders helps to define the "rooms" within an outdoor environment.

Four concrete finishes side by side illustrate the diversity of this amazing material.

A random stone pattern in stenciled concrete creates an attractive foil to a fountain centerpiece.

Color variations add to the authentic feel of this "terracotta" terrace.

Concrete, stained to resemble a mix of flagstones, hugs the back of this contemporary-style home.

Darker "paving stones" break up the basket weave pattern of a patio and walkway.

A flowerbed is a natural centerpiece on a small patio.

Stamped concrete was introduced to the West Coast in the 1960s. Today it has spread across the United States and all over the world.

Another benefit of using concrete for decks and patios is that tables and chairs will not rock or wobble on a surface that is generally smoother than stone or brick.

A raised patio with "stone" surfacing carries with it an air of permanence.

Not just for private homes, a stamped concrete patio can find a place just about anywhere, even in a corporate setting as this patio outside an office building's entrance proves.

Using stamped concrete, a builder added character to a church patio. The random pattern was chosen to complement the style of the church's exterior walls.

A surface of imitation stone was designed to ramble about this backyard, creating a sense of spontaneity.

Dark concrete borders frame the short retaining walls around flowerbeds.

To give the backyard décor a sense of unity, the patterns used on the back porch continue out to the pool deck and dining area.

Ashlar "slate" complements the rustic look and feel of this country retreat.

In keeping with the overall design, the countertop on the half-wall is also made of concrete and has been stained to match the patio.

A European fan pattern supplies visual texture, creating a classic look.

The curved shape of the European Fan pattern lends itself to this patio's half-circle shape.

Concrete was cast to fit the slope and layout of a backyard, making it seem a natural extension of the home.

Concrete squares and rectangles in an arresting series of patterns play across the surface of this backyard.

An ashlar pattern surface complements the classic look of pillars.

Brick half walls border a stamped concrete patio.

The texture of stamped concrete adds visual interest to this shady passageway.

Using a common pattern on multiple outdoor surfaces lends unity to spaces that have different shapes, positions, and functions.

Scale is important in planning a patio, making it proportionate with the surrounding architecture and landscape.

A matching curb edges a basket-weave "paving stone" patio.

A rosy sandstone texture creates the right atmosphere for a tropical retreat.

Positioned just right, a patio takes full advantage of the view and breeze from a nearby inlet, the pattern on the patio is reminiscent of old fashioned garden paths paved with a jumble of loose stones.

Sandstone textured concrete adds rustic charm to an upscale patio setting, complete with firepit.

Instead of using walls, different colors and patterns create dividers free of physical barriers for this outdoor environment.

Just like rugs can add color and decoration to the interior room, stamped and stained concrete can do the same for an exterior room.

With concrete rocks, a waterfall, and cactus and other desert plants growing in the flowerbeds, this patio seeks to appear similar to the desert landscape behind it. The rustic texture of the patio surface complements the overall surroundings. Sandstone texture and tone blend with the adobe home.

(Continued on following 2 pages)

A paved balcony offers an excellent view of the matching stamped concrete patio below.

While elevated and separate, a common stamp pattern unites a dining pavilion with the main patio.

Ashlar slate covers the porch in front of a small side entrance.

A basket weave pattern adds a tropical touch to this wicker-furnished patio.

By individually staining the "stones" within a stamped pattern, a convincing effect of real masonry is accomplished.

Modern concrete stamping tools can reproduce a wide variety of textures, including termite-proof wood.

A herringbone pattern has been "snazzed up" and carefully keyed to the home with the application of various color agents to the "paving stones" capping a raised deck. The same colors are repeated on the retaining walls below, this time in a random stone pattern.

Darker accents on the patio surface give it an elegant, weathered look.

While concrete can be stained in a wide variety of colors, most homeowners gravitate toward colors similar to native stone of their regions.

A curved wall creates a private outdoor niche for this pretty little patio.

Two marble-like finishes add classical appeal to this porch environment.

Built-in bench seating in concrete overlooks a brick-red concrete patio, the two finishes nicely complementing each other.

Concrete takes on the form of terra-cotta tiles for a tropical patio room.

A close-up reveals how stamping and staining can be carefully accomplished to span corners and edges.

For added decoration, parts of an ashlar pattern have been stained different shades of slate gray.

A raised flowerbed provides a lovely introduction to this home, as do curved curbs that usher a visitor on the winding path to the front door.

If one did not look carefully for the seam running across the patio, the casual observer would have no idea that this "flagstone" patio is stamped concrete.

Brick fans spread out from a smooth circle in the center of the driveway, creating a ripple effect.

A massive crest has been stained onto the surface on this courtyard. Four lines of yellow tiles radiate out from the center, further drawing attention to the crest.

Variations in color help break up a broad plaza into zones.

Colorful lines spread out from the center and fill this space with a dizzying pattern of ribbon-like stripes.

The European fan pattern originated from a pattern established when men would set stones in a semi-circle the length of their arms, then move to the next location and start again.

Shot from above, a carefully planned courtyard combines three tones of concrete, contrasting with green corners, for a luscious centerpiece from any level.

Poured concrete takes on the look of timeworn cobblestones on a broad plaza.

Three tones were married for an abstract work of concrete art.

Stencils and powdered colored hardeners were used to create a mosaic-like tile border around a patio surface, with a mottled finish to mimic quarried stone.

Pools

The same slate imprint and stain used on the deck surface were continued on the walls of a raised flowerbed, and brick was mimicked with eye-fooling artistry.

Three concrete treatments side-by-side combine design, color, and texture for rich effect.

A concrete pool edge was carefully sculpted to create an illusion of real rock skirting.

Here, a children's slide has been built into the shape of a tugboat. Concrete rocks installed around the pool create a grotto effect. An exposed aggregate surface adds grip and safety to the patio/play area near the pool.

A cluster of "rocks" simulates a natural spring or waterfall.

One of the more popular concrete stamp patterns, ashlar slate surrounds this pool setting.

A natural looking deck complements the waterfall and grotto at the opposite end of the pool.

Natural looking concrete stones flank a pool, and provide décor, seating, and a sense of a natural setting.

Colorings have been varied for this low-maintenance backyard environment, helping to define different areas and coordinating with the abstract patterns on the pool's bottom.

Concrete surfacing on both the faux stone patio pool surround and the stucco walls of outdoor showers has been stained in the rich, golden-red hues of terracotta.

A concrete dam and pump house are faced in faux stone.

Colored concrete helps break up and establish zones within a large pool complex.

A detail shot reveals the eye and toe-fooling effect of stamped concrete.

The surface of the pool deck was given a light chemical stain in order to help give the area an antique look that fit in with the architecture of the main house.

The free-form shaped edge of the pool, a flagstone path, and a faux-stone patio create the illusion that one has discovered a natural oasis.

White skirting around the spa and planter beds adds punctuation to the expanse of "flagstone" surround, and coordinates with the privacy walls and home beyond.

White surfacing around a pool reflects light, and remains cooler for bare feet. In climates where the sun tends to be harsh, a darker surface can reduce glare, making it easier on the eyes.

Concrete surfacing was stained to fit the southwestern setting of a pool and patio.

Random stone patterns are very popular with builders and homeowners.

This pool features two waterfalls, one filling the Jacuzzi and one filling the main pool. The concrete rocks surrounding each water feature have been stained different colors than the two-toned patio.

Various angles reveal a simple rectangular pool engulfed in luxury. Amidst a unifying stamped concrete patio, fountains play, a spa gurgles, and a fireplace glows atop a raised surface.

An ashlar pattern consists of rectangular stone blocks with square edges and even faces; it is one of several patterns commonly reproduced by stamping concrete.

There are a variety of ashlar patterns used in decorating concrete, with the difference being the size of the blocks, their position, and the kind of stone being reproduced.

A gray pool cover blends in with the ashlar slate pattern stamped on the pool deck. A recent rainstorm has given both a glossy quality.

The surface of the pool deck and the surface of the patio have been stamped with the same pattern, giving both spaces a dignified and united look.

The pool deck's uneven edge combines with the exposed aggregate on the pool surround to give the area a more natural look.

A natural pattern works well to create a more relaxed, casual environment.

Imitating a cobblestone road, this concrete pool deck was stamped with a pattern of large flagstones surrounded by smaller cobbles.

Fan-shaped steps descend to a natural looking swimming pool surrounded by rocks, flowers, and trees.

Another example of an ashlar pattern, note that the color of the pool deck is the same as the color of the pool surround with a river stone border.

Because it is possible to split slate so thin, it is frequently used for roofing and floor tiles. As a result, slate patterns are very popular with homeowners who want to achieve the same look with concrete.

A rock garden at the edge of the pool hides the machinery needed to make a waterfall, making it look like the pool is fed by a natural spring.

Jutting out into the pool, a flowerbed takes on the appearance of a tropical island surrounded by the cool tropical ocean.

A house seems to end at the very edge of a pool, thanks to a columned deck that shelters the underlying patio.

Concrete blocks, set into the pool to look like stepping stones, guide visitors across the pool to the bubbling waters of a hot tub. A cleverly positioned statue of a mermaid suits the nautical décor while keeping people from walking on the serving counter of a sunken bar.

The cool color and texture of slate complement the peaceful atmosphere of this backyard oasis.

After dinner on the patio, guests can move down to the pool deck, which has been stamped in a pattern that complements not only the patio, but the décor of the house as well.

Here a swimming pool becomes a welcome oasis amidst a profusion of tropical plants and flowers.

In the fading afternoon light, the surface of this pool deck takes on pinkish hues. A random stone pattern complements the rocks set around the pool's edge.

A contemporary effect is created by the straight lines of "cut stone" edging this pool.

A raised white platform adds reflected light for sun worshipers, while warmer tones welcome bare feet around the edges of the pool.

An interesting feature of this pool is a fitted cover that slides neatly under the coping.

The concrete on the pool deck was stained to look like sandstone, complementing the Southwestern roof tiles on the house.

Sections of red "brick" add accent and color to the surrounding pool deck.

With the addition of some palm trees, a waterfall, and a grotto, this pool is transformed into a tropical paradise. The random slate pattern underfoot adds to the sense of Eden.

Karchi Photography

An artful hand created the design for pool and deck, creating an appealing picture from every window that overlooks this masterpiece.

A shield-shaped spa is separated by a divider capped in the same decorative concrete as the pool deck.

Individual "stones" within a stamped pattern were carefully hand-colored to create this natural effect.

An interesting effect was created by interrupting the white concrete skirt circling the pool with strips of "flagstone."

Stamped concrete can reproduce the look and feel of a terracotta surface.

Semi-circular "stone" decks invite sunbathers close to the edge.

Brown and tan "tiles" mix with an awning to create an atmosphere reminiscent of an old Roman garden.

A closer look at the pool deck reveals the detail that went into stamping, staining, and texturing the concrete.

Glass block, blue tile, and concrete "stone" combine for a colorful pool environment.

The lack of borders or linings gives the impression that this flowerbed formed in the natural break in the rock.

Neighboring pools reflect a diversity of tastes. Though similar surfacing materials were put to work, the pool in the foreground was designed for a lap swimmer and sunbathers.

A square, stamped surround is complemented by a smooth, curving pool edge stained to match.

Like a river running through it, a colorful, wavelike mosaic pattern ripples the length of a pool area.

Conforming to the environment around it, this walkway seems to flow down the hill.

A swath of red makes a division between two sections of a public walkway, drawing attention to the different colors and patterns of the two sections.

To provide traction and keep people from slipping, treads were etched into the edges of these steps.